The Amazing World of
Microlife
That Lives in Soil

Steve Parker

⬤ Raintree

Chicago, Illinois

Chicago, IL

For information, address the publisher:
Raintree, 100 N. LaSalle, Suite 1200
Chicago, IL 60602
Customer Service 888-363-4266
Visit our website at www.raintreelibrary.com

Printed and bound in China by South China Printing Company Ltd
10 09 08 07
10 9 8 7 6 5 4 3 2

Edited by Katie Orchard
Designed by Tim Mayer
Picture Research by Lynda Lines and Frances Bailey
Production by Duncan Gilbert

Library of Congress Cataloging-in-Publication Data
Parker, Steve.
  Microlife that lives in soil / Steve Parker
     p. cm. -- (The amazing world of microlife)
  Includes bibliographical references and index.
  ISBN 1-4109-1846-7 (lib. bdg. : hardcover) --
  ISBN 1-4109-1851-3 (pbk.)
  ISBN 978-1-4109-1846-8 (lib. bdg. : hardcover) --
  ISBN 978-1-4109-1851-2 (pbk.)
1. Soil microbiology--Juvenile literature. I. Title.
II. Series: Parker, Steve. Amazing world of microlife

QR111.P329 2005
579'.1757--dc22

2005004228

Acknowledgments

The publishers would like to thank the following for permission to reproduce photographs: Alamy p. **24** (Nigel Cattlin/Holt Studios Int. Ltd.); Corbis pp. **28** (Robert Pickett), **29** (Michael and Patricia Fogden); Getty Images p. **5 top** (David Wolley/Stone); photolibrary.com pp. **5 bottom** (David M. Dennis), **10**; Rex Features p. **15** (Marja Airio); Science Photo Library pp. **1** (VVG), **3** (VVG), **4** (Andrew Syred), **8** (David Scharf), **11** (Dr. Jeremy Burgess), **12** (Dr. David Patterson), **14** (Eye of Science), **16** (Laguna Design), **17** (Andrew Syred), **18** (VVG), **19** (Andrew Syred), **21** (VVG), **22** (Geoff Kidd), **26** (Andrew Syred), **27** (Peter Chadwick); Still Pictures pp. **7** (Kelvin Aitken), **25** (Gayo); Topfoto pp. **6** (T. Balabhadkan/UNEP) **9** (Bob Daemmrich/The Image Works), **13** (Bob Daemmrich/The Image Works), **20** (Roberto Matassa).

Cover photograph of a velvet mite reproduced with permission of Science Photo Library (Eye of Science).

Every effort has been made to contact copyright holders of any material reproduced in this book. Any omissions will be rectified in subsequent printings if notice is given to the publishers.

The paper used to print this book comes from sustainable resources.

Contents

Some words are shown in bold, **like this**. You can find out what they mean by looking in the Glossary.

Microjungle!

Every time you visit a garden or park, you walk over a microjungle! The soil under your feet is wriggling with living things. We can easily see some of them, such as earthworms and **grubs**. Others are too small for our own eyes to see. But we can make them look much bigger, using magnifying glasses and **microscopes**.

A microscope can make this mite look as big as your hand. In real life, it is only as big as this period.

A magnifying glass makes minibeasts like this earthworm look much bigger than they really are.

Why we need soil

Soil with plenty of microlife in it is rich, or **fertile**. It has everything a plant needs to stay healthy and grow. Grass, vegetables, fruit, and crops such as wheat are all plants. If there was no microjungle in the soil, plants could not grow, and we would be very hungry!

Rich and Poor Soil

There are many kinds of soil. Rich, or **fertile**, soil has plenty of plants growing in it. Grass and trees grow well because the soil is home to a lot of microlife. All living things need water, food, and fresh air to help them grow. Fertile soil with plenty of air holes in it has all these things, and so is full of life.

Good and bad

Other kinds of soil are not fertile. They are poor soils, which have little microlife and few plants. Soils become poor because they lack water and are too dry. They may have few **nutrients** as food. Some soil is too squashed together, with no air spaces in it, so plant roots cannot grow there.

If soil becomes very dry and cracked, the microlife that lives there will die.

Good soil is damp and crumbly. It has plenty of tiny spaces for air and water, and a lot of nutrients.

WHAT IS HUMUS?
Rich soil has plenty of **humus**—bits and pieces of dead plants and animals. They rot and break up, slowly turning into new soil. Humus provides a lot of nutrients, which microlife and plants use as food.

Microbugs

The tiniest living things in soil are called **bacteria**. One handful of soil contains millions of bacteria. Some soil bacteria grow like plants, using light from the Sun to make their own food. They can live only near the surface, where there is enough light. Other kinds of bacteria do not need light. They feed on the **nutrients** in soil.

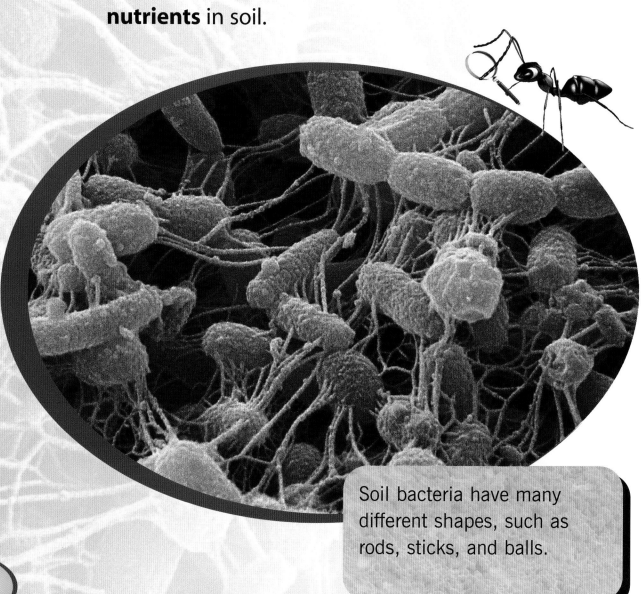

Soil bacteria have many different shapes, such as rods, sticks, and balls.

Gardening is fun. But we should cover cuts and scrapes safely, to stop harmful bacteria from the soil getting into our bodies.

How bacteria help

Bacteria help make soil rich and **fertile**. They help to rot (break down) **humus**. This provides food for larger plants and animals. Bacteria are also eaten by tiny animals in the soil.

CARE WITH SOIL
Some kinds of soil bacteria can be harmful and cause serious illness. We should never put soil in our mouths, and we should always wash our hands thoroughly after touching soil.

Microplants

Many kinds of plants, such as shrubs, flowers, and trees, grow up from the soil. Other plants grow inside the soil, too. Most of these plants are so tiny that they look like specks of green powder or even green slime!

Algae are tiny plants that live in soil. They do not have roots, stems, and flowers, like big plants. They are mostly shaped like tiny green pieces of string.

Light and water

Just like big plants, microplants need light from the Sun. They live on the surface of the soil or just underneath, where the light can reach them. These microplants also need water. They grow in the very thin layer of water that covers the tiny grains of soil. Microplants in soil are eaten by all kinds of tiny creatures. These tiny creatures are then eaten by bigger ones, and so on.

As these puddles dry up, they leave microplants around their edges, which look like colored powder or slime.

SMALL AND BIG
Most microplants in soil are called algae. Much bigger kinds of algae grow on the seashore—we call these large algae seaweed.

Sliding Along

What wobbles like Jell-O as it moves through soil? The answer is an **amoeba**. The amoeba belongs to a group of tiny living things called **protists**. The amoeba eats smaller microlife such as **bacteria**.

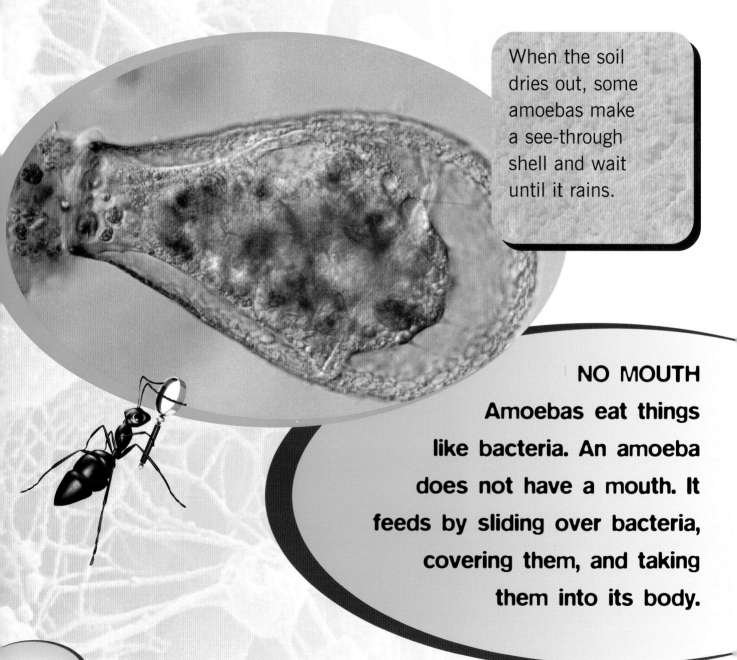

When the soil dries out, some amoebas make a see-through shell and wait until it rains.

NO MOUTH
Amoebas eat things like bacteria. An amoeba does not have a mouth. It feeds by sliding over bacteria, covering them, and taking them into its body.

Other protists

There are many other kinds of soil protists, too. Some are shaped like sausages or cushions, with a long thread sticking out. A protist waves this thread around, to push itself through the soil.

There are millions of protists in a bucket of soil. Some are green and grow like plants, using sunlight to make their food. All of these protists are food for slightly bigger hunters in the soil.

If soil gets too dry, the microlife will die. There will not be enough food for the bigger plants and animals, and they will die, too. This is why we sometimes need to water our gardens.

Recycling Soil

In damp weather, mushrooms and toadstools grow on soil. These kinds of living things are called **fungi**. Fungi cause rot and **decay**. They break dead plants and animals into tiny pieces, which crumble and soak back into the soil. Fungi grow tiny threads into the soil, which turn old bits of plants and animals into a "soup." The fungi then soak up the soup as their food.

The tiny threads of a fungus form a tangled net in the soil.

Mushrooms and toadstools grow up from much smaller threads hidden in the soil.

How fungi spread

Threads of fungi sometimes bunch together and grow upward, to form a mushroom or toadstool. The mushroom's top, or cap, makes millions of powder-like specks called **spores**. Spores float away in the wind. If they land in a suitable, damp place, they grow into a new fungus.

NATURAL RECYCLING

Fungi are nature's way of recycling. If they did not cause rot and decay, dead bits of plants and animals would pile up and soon cover us!

Tiny Creatures

What's the smallest animal in soil? You may think of a baby earthworm, but some creatures are much tinier. One is called a rotifer. It has a mouth at one end, with tiny hairs around it, in a ring or wheel shape. Another tiny soil creature is the water bear. As it walks along on its stumpy legs, it looks like a tiny bear.

A rotifer has a tiny body shaped like a vase or funnel, with no legs.

BACK TO LIFE

Rotifers and water bears live in the water in soil. If the soil dries out, they can still survive. When the soil gets wet again, they come back to life.

Microhunters

Rotifers and water bears live in the very thin layer of water around pieces of soil. Rotifers eat anything smaller than themselves, such as microplants and microbugs. Water bears suck the juices out of plants. Both of these tiny creatures are then eaten by bigger soil animals.

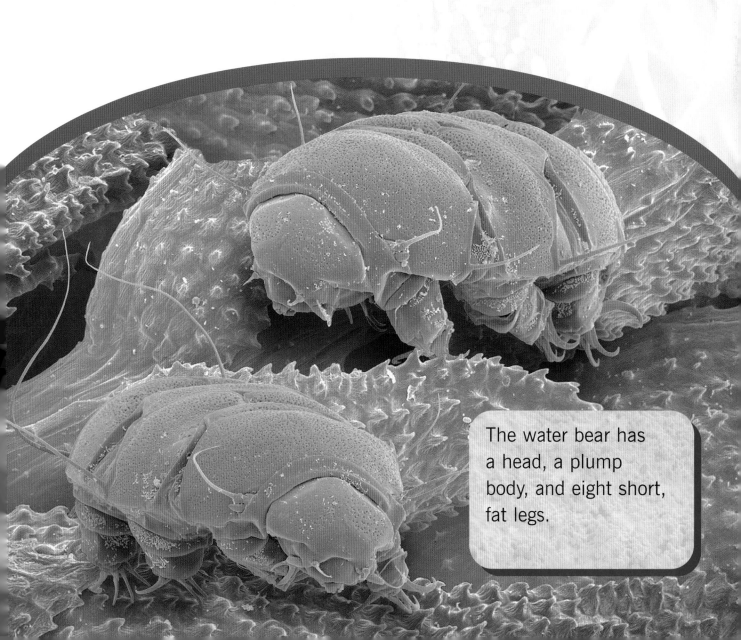

The water bear has a head, a plump body, and eight short, fat legs.

Soil Mites

If we look closely at the ground, we may see a small spider running across the soil. We know it is a spider because it has eight legs. The spider could be hunting for an even smaller creature, which also has eight legs. This is the type of tiny soil animal called a mite.

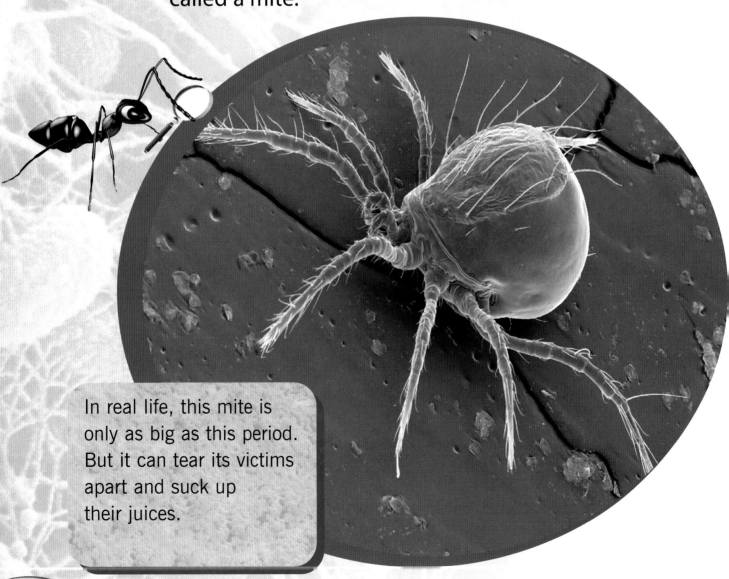

In real life, this mite is only as big as this period. But it can tear its victims apart and suck up their juices.

Small and smaller

There are many different kinds of soil mites. Some are as big as this "o." Others are a hundred times smaller. The tiniest kinds eat whatever they find, mainly bits of dead plants and animals. Bigger mites have strong mouths shaped like pincers. They hunt all kinds of smaller creatures, including other mites.

FEELING THE WAY
Many soil mites have no eyes. Under the ground, there is no light, so eyes are useless. These mites touch and feel their way through the dark soil.

fangs

One of the mite's main enemies is the centipede. It stabs its live food, or **prey**, with its very long, toothlike **fangs**.

Pincers and Springs

In a scoop of soil, you can easily see creatures like earthworms. Look more closely and you might see little specks moving. These are tiny animals that live in the soil. One type is the false scorpion. It has pincers like a real scorpion, and eight legs. Another type is the springtail. It has a head with eyes and two bendy feelers, and a long body with six legs.

NO STING
A false scorpion looks like a tiny version of a real scorpion—except that it has no tail, and so it cannot sting.

Many tiny creatures hide in **leaf litter**— the layer of old leaves, twigs, and other bits and pieces on top of soil.

Under the springtail's tail is a Y-shaped part that can flick down very fast, to make the springtail leap into the air.

Different lifestyles

Springtails and false scorpions live in very different ways. Most springtails eat any small bits and pieces of food they can find. Most false scorpions are fierce hunters. But in the soil, all of them become food for larger creatures.

Grubby Grubs

Many kinds of creatures spend all their lives underground. But some stay in the soil only while they are young. Many are the young of insects, called **grubs**. The soil is their nursery. As they change into grown-up insects, they come to the surface and run or fly away!

STAGES IN GROWING UP

The grub or young form of an insect is called a **larva**. It wriggles and feeds. Then it grows a hard case and becomes a **pupa**, which lies still. Then the pupa hatches into the grown-up insect.

The leatherjacket lives in the soil for a year or more, before it changes into the grown-up crane fly.

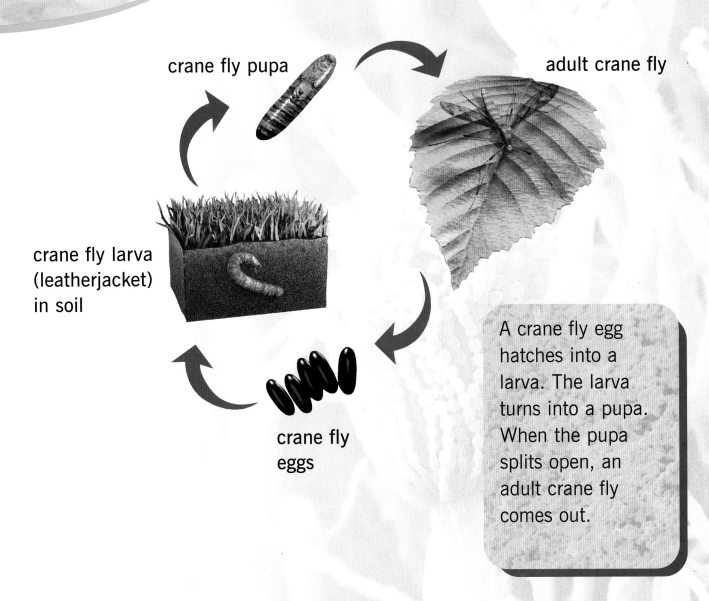

crane fly pupa

adult crane fly

crane fly larva
(leatherjacket)
in soil

crane fly
eggs

A crane fly egg
hatches into a
larva. The larva
turns into a pupa.
When the pupa
splits open, an
adult crane fly
comes out.

Grubs to grown-ups

One of these grubs is the leatherjacket. It is brown and it
eats plant roots. It wriggles like a **maggot** because it has
no legs. After a time, it grows a hard case, or "jacket,"
around itself. When the case splits open, a crane fly
comes out and flies away across the soil.

Soil Pests

In wild areas, there are many different kinds of plants and animals. But in farming, the land is used to raise just one type of crop. A huge field of one kind of plant, such as potatoes, is like a giant feast for certain kinds of soil creatures. These creatures feed well and breed faster than they would in a natural area. They soon become **pests**.

This young, or **maggot**, of the cabbage root fly is hiding in a plant stem. Millions of these maggots can soon destroy a field of cabbages.

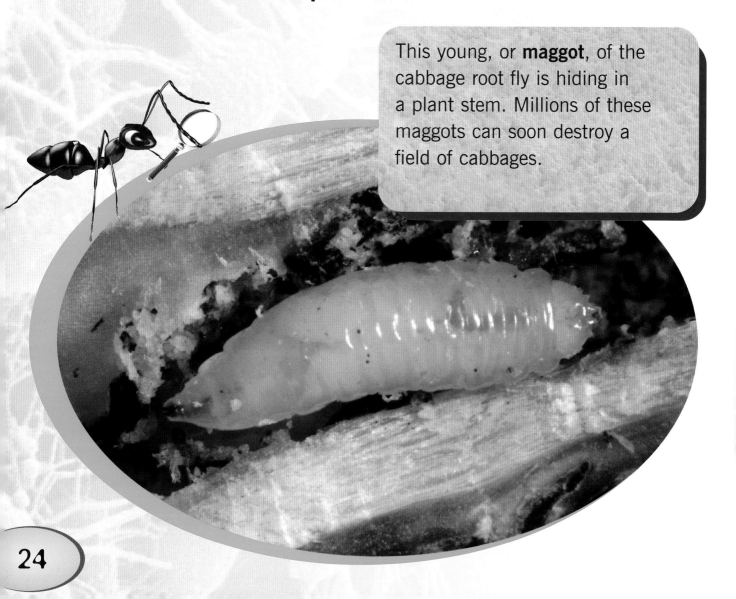

Too much food

One type of pest is the wireworm—the young of the June beetle. It nibbles crops such as wheat. Various kinds of mites can become pests, too, as they eat vegetables growing in the soil. Eelworms also eat their way into vegetables and plant roots. These tiny soil creatures can ruin a field of crops in weeks.

SOIL PESTS
Around the world, about one-tenth of all the food that farmers grow in fields is eaten or ruined by animal pests.

Farmers often spray their crops with chemicals to kill pests.

Underground Cities

Big cities can be very busy. There are always many people rushing around. The places where minibeasts such as ants and termites live in the soil are like cities, too. These tiny insects dig tunnels all day and night. They move grains of soil with their pincer-like mouths. Thousands of them work together to make a nest of burrows and chambers in the soil.

BIGGEST CITIES
Termites make the biggest cities of any animal. There might be more than 10 million termites in one huge nest.

In a busy nest, ants "talk" to each other by tapping their feelers together.

Always working

One special ant or termite in each underground city is the queen, who lays eggs. The rest are workers. Some make new tunnels and repair old ones, while others gather food. Some keep the eggs clean and feed the young. There are soldiers, too. Soldier ants and termites fight enemies, from small ant-eating spiders to the huge anteater!

Termites build a huge mound or tower of hard, dry mud above their nest in the cool, damp soil.

Nature's Gardeners

Almost every patch of soil has long, thin, and wriggly worms. Some of them are quite large, and can be longer than your finger. These are earthworms. Other worms are too small to see. They are called roundworms.

Some worms, such as this common earthworm, eat soil. Others feed on tiny creatures or suck juices from plant roots.

NATURE'S GARDENERS

Earthworms are sometimes called "nature's gardeners." As they tunnel, they mix up the soil and spread out the **nutrients**. It is like a gardener digging soil—although much slower!

Tasty soil!

Worms do not just live in soil, they eat it! The worm takes in the nutrients from the soil. Then it gets rid of what it does not need through its tail end, as squiggles of soft mud, called worm casts. As worms eat, they make tunnels, which let light, air, and water into the soil. Worms also pull into their tunnels old leaves and bits of plants that rot and **decay**. Worms are very good for soil.

Millipedes feed on old and dead leaves. Sometimes they hide from their enemies in worm tunnels.

29

Find Out for Yourself

More Books to Read

De Bourgoing, Pascale, and Daniele Bour. *First Discovery: Under the Ground*. New York, N.Y.: Scholastic, 1995.

Fredericks, Anthony D., and Jennifer Dirubbio. *Under One Rock: Bugs, Slugs, and Other Ughs*. Nevada, Nev.: Dawn Publications, 2001.

Himmelman, John. *Nature Up Close: An Earthworm's Life*. Danbury, Conn.: Children's Press, 2001.

Tomecek, Steve, and Nancy Woodman. *Jump Into Science: Dirt*. San Francisco, Calif.: National Geographic Children's, 2002.

Using the Internet

Explore the Internet to find out more about microlife that lives in soil. Use a search engine and type in a keyword such as fungus, mite, rotifer, or humus, or the name of a particular type of microlife.

Glossary

algae plants that have no flowers or roots. They range from huge seaweed to tiny, green, hairlike types.

amoeba type of protist, like a tiny bag of Jell-O

bacteria tiny living things. Some bacteria are helpful and some cause disease.

decay break apart and become rotten or moldy

fang long, sharp tooth

fertile containing lots of nutrients to help living things grow

fungi group of living things including mushrooms, toadstools, and yeasts, which cause rotting or decay

grub kind of small, wriggly, wormlike creature

humus brown, powdery parts of soil made by decay. Humus is rich in nutrients.

larva young form of a small creature such as an insect, which hatches from an egg

leaf litter mixture of dead leaves, bits of twigs, and loose soil found under trees

maggot legless young or larva of a fly

microscope equipment to make very small things look bigger

nutrients substances used by living things to grow and stay healthy

pest living thing that causes damage to people, their homes, their animals, or their plants

prey creature that is hunted by another one for food

protist type of tiny living thing. Many look like specks of Jell-O.

pupa stage in an insect's life when it grows a hard body case and stays still, before it changes into an adult

spores tiny, seedlike parts made by fungi, which grow into new fungi

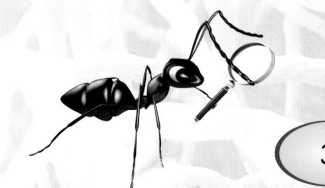

Index